Beach Trip Biology–
An Educational Snorkeling Guide to the Caribbean

Dawn Lamb

Copyright © 2013 Suitcase Studies, LLC.

All rights reserved.

ISBN – 10: **1491279389**
ISBN–13: **978-1491279380**

CONTENTS

	Introduction	4
1	Snorkeling Basics	5
2	Types of Reefs	9
3	Coral Reef Energy Transfer from the Sun	11
4	Symbiotic Relationships on the Reef	15
5	Biological Classification	18
6	Vertebrates on The Reef – Anatomy	26
7	Vertebrates on The Reef – Behavior	33
8	Common Vertebrates on the Reef – Identification	35
9	Invertebrates on the Reef	44
10	Plants on the Reef	66

INTRODUCTION

Beach Trip Biology is meant to be a guide to provide children with learning activities centered around Biology while making a trip to the Caribbean or South Florida. No matter the number of days traveled, this guide can provide some background knowledge so that children can develop some of the basic concepts of biology around the travel experience.

The book is not intended to be a science book, but a guide to develop background knowledge so that educational experiences can be built while traveling in the Caribbean or South Florida.

Much of the activities referred to in the book refer to snorkeling. Snorkeling is an outdoor activity with a certain degree of risk. All activities should take place in the presence of adults who understand the risk and supervise all activity.

All of the photos used in the book (unless noted) are taken by the author while traveling to St. Thomas, US Virgin Islands or Roatan, Bay Islands, Honduras. The author is not a professional photographer but an educator who provides the photos as examples of typical observations during trips.

1 SNORKELING BASICS

Do not harass or touch the plant or animal species on the reef

It is not appropriate to disturb the species that are living in the coral reef ecosystem. It may be tempting to try to touch the fish or coral, however, the oils from skin will damage the protective cells of the fish and coral and jeopardize its health. Additionally, do not attempt to chase or otherwise disturb the fish. Snorkel 'quietly'– making as little disruption as possible and quietly floating above the reef. A quiet snorkeler will see more natural behaviors of the species by creating as little disruption as possible. Avoiding sudden movement will result in the reward of greater observation of a variety of species.

Avoid all contact with the coral

When venturing through the reef, a snorkeler will face times when the reef is closer to the surface than in other places. While attempting to move across the coral, be sure that the water is deep enough that it can be swam over without making contact with the coral. If the water is too shallow to swim across, a snorkeler should choose another swimming route.

Never stand on coral. If a situation occurs where it is necessary to touch down (to adjust a mask or other equipment), swim to a sandy or grassy area.

Avoid leaving anything on the reef

Do not carry products which may dissolve or drop onto the reef. Do not leave trash or any unnatural substances in the water. Additionally, when trash or unnatural objects are seen in the reef area, pick them up (if it can be done without disturbing other species) and carry them to shore.

Do not feed the fish!

It is not appropriate to feed the fish. Tourists often engage in feeding the fish so that the fish gather around in large groups. The problems are that:

1) The food is not good for the fish. It is not in their natural diet.
2) The fish begin to rely on the human distribution of food and forget how to forage for their own food. When the humans leave, the fish don't know how to feed themselves.
3) The balance of the ecosystem is disturbed. The natural food supply of the fish goes 'unchecked' and therefore multiplies beyond a normal level. As a result, the population of the species that the fish would otherwise be eating grows larger. As this species' population gets larger, stress is placed on the ecosystem for its food source and the 'ripple effect' continues.

Even when the practice is encouraged by locals or tour guides, it is not healthy for the fish and should be avoided.

Be careful of sunburn

Be mindful of the exposure to the sun. While floating on the frontside, a snorkeler's back is getting cooked. Because the back and legs are being cooled by the water, a snorkeler may not realize how much sun exposure is taking place. Use appropriate sunscreen and reapply frequently! It is advised to wear a t-shirt or rash guard, particularly after a couple of days of snorkeling and exposure to the sun.

Equipment

Mask – Snorkelers will need a mask that fits snuggly. A basic mask is sufficient for most snorkeling activity. It is important that the mask fits so that water does not enter the mask.

Snorkel – A basic snorkel is all that is necessary. It is helpful if the snorkel has a wave deflector or float which helps to keep the water out of the top of the snorkel if a wave crashes over the snorkeler. Most coral reef regions are relatively calm, so the wave deflector is not essential.

Water Shoes / Fins – Most snorkelers choose not to use fins while snorkeling on the coral reefs because they tend to 'kick up' sand, sometimes limiting the visibility while snorkeling. In that case, water or swim shoes should be worn to protect feet and toes from rocks. However, if it is necessary to swim a distance out to the reef or if the water is deep and the snorkeler wants to swim to the bottom for closer observation, fins may be helpful for snorkelers who are not strong swimmers. If fins are worn, the snorkeler should be very mindful to avoid the fins making contact with the coral.

Snorkeler vest – For snorkelers who are not strong swimmers, a snorkeler vest may be helpful. The vest is similar to a life vest but it has adjustable buoyancy that is adjusted by the swimmer by inflating the vest. The snorkeler vest allows the swimmer to float on top of the water with very little effort. Snorkel vests are not always available at dive shops and if it is desired, it would be suggested that one is obtained prior to going to the islands.

Mask clearing solution – Masks will fog under most circumstances. Clearing solutions can be purchased at most dive shops, however, a simple solution of baby shampoo will work as well and is much less expensive. Put a couple of drops on the lens of the mask and rub it around. Let the mask 'air out' some to take away some of the 'fumes'. The solution will 'sting' eyes for just a few minutes if it is not 'aired it out' sufficiently. Some snorkelers clear masks with saliva. This will help the mask stay clear but is usually not sufficient for long snorkeling periods.

Check for Understanding:
1) Why should snorkelers avoid feeding the fish?
2) Why is it important to not touch the coral?
3) When should snorkelers consider wearing fins?

2 THE TYPES OF REEFS

> **Lesson Objectives**
> 1. Students will be able to identify the different types of reefs and the characteristics of each.

Atoll – An atoll is a coral reef that is enclosed, at least partially, and has a deep lagoon in the center. Atolls are typically created around a volcanic mountain in the ocean. Sometimes the volcanic mountain has sunk below the surface of the ocean and the reef remains alone.

(photo from Wikipedia - public domain)

Fringing Reef– A fringing reef grows off the coast of a tropical landmass. The illustration below shows the parts of the fringing reef.

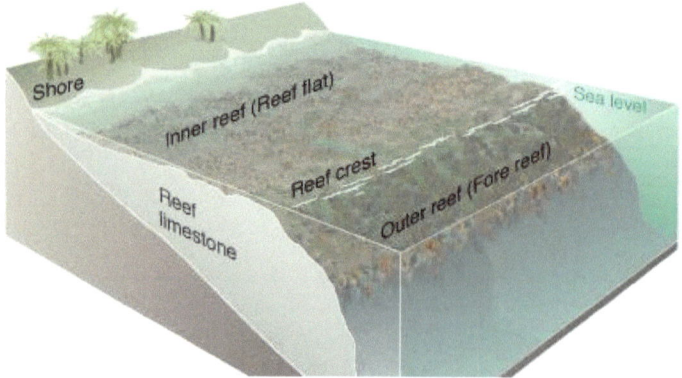

(illustration from Wikipedia - public domain)

Barrier Reef– A barrier reef grows off the coast of a tropical landmass but differs from the fringing reef due to the deep lagoon between the reef and the shore. Barrier reefs are off the shore and tend to run parallel to the shoreline.

Check for Understanding:
1) What are the different types of reefs?
2) Be able to describe the different types of reefs.

3 CORAL REEF ENERGY TRANSFER FROM THE SUN

> ### Lesson Objectives
> 1. Students will understand the various species types and how they 'fit' into the food chain of the reef.
> 2. Students will be able to explain how the sun provides all living organisms on Earth with energy.

Biology Essential Concept:

The source of all energy on Earth originates from the sun.

All energy on Earth comes from our sun and **food chains** are representations of the sequences of organisms related by which organism feeds off the others. Every ecosystem has specific organization related to how energy is transferred from one species to another. While the sun is the ultimate source of all energy on earth, the energy from the sun is transferred to other organism as they eat.

Solar Energy Transfer

All energy on earth originates with the sun. All living organisms depend on energy to function. How that energy is transferred from the sun to all of the organisms that live on earth is represented in food chains. A food chain represents the transfer of the solar energy from one organism to another.

Photosynthesis

The most fundamental of the energy transfer from the sun comes from **photosynthesis**. Photosynthesis is the process by which plants convert the sun's energy into sugar which is used by the plants to grow and function. Photosynthesis (photo = light, synthesis= make) is a basic requirement for life on earth. Organisms which produce energy from photosynthesis are referred to as **producers**.

> **Vocabulary**
> **consumers** – organisms which eat other organisms
> **food chain** – representations of the sequences of consumers and producers that transfer the sun's energy
> **photosynthesis**– the process that is used by producers to make food from the sun
> **primary consumers** – first level organisms in the food chain that eat producers
> **producer** – organisms which make their own food from the sun's energy

Producers

Producers transfer the energy from the sun to other organisms when they are digested. So, when a cow eats grass, for example, the energy from the sun that was produced by the grass is transferred to the cow so that it has energy to function. On the coral reef, algae and other ocean plants are the producers in the ecosystem. As a result, they are at the very bottom of the **food chain**.

Primary Consumers

Consumers are defined as those organisms which 'eat' or consume producers or other consumers. **Primary consumers** are those organisms which eat the producers.

Secondary Consumers

Secondary consumers are those animals which eat primary consumers. They are one 'level' up the food chain from primary consumers and two levels 'up' from producers.

Tertiary Consumers

Tertiary consumers are those animals which eat secondary consumers.

Food Webs

Food webs represent the complexity of the producers and consumers in an ecosystem. Where food chains are 'one-directional', food webs are more complex and show the different relationship of a variety of species and how the energy transfer is made throughout the ecosystem.

Food Chain

Algae ⟶ Parrotfish ⟶ Barracuda ⟶ Shark

Food Web (examples)

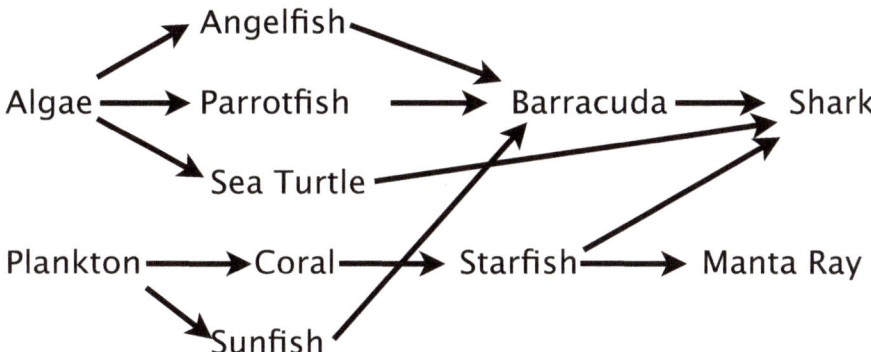

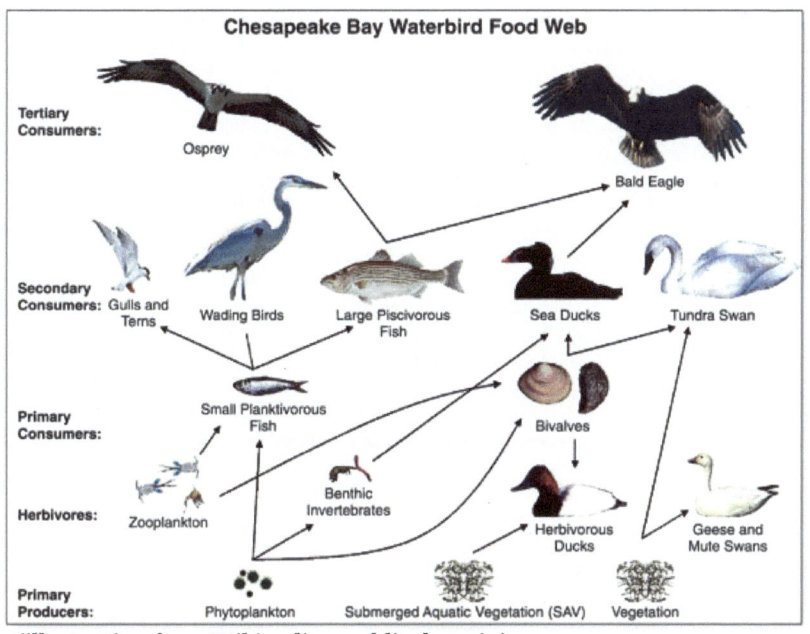

(illustration from Wikipedia - public domain)

Check for Understanding:

1) Describe how the energy from the sun is transferred to various animals in an ecosystem.
2) What is the difference between a producer and a consumer?
3) Describe the difference between a food web and a food chain.
4) What is the difference between a primary consumer and a secondary consumer?
5) Explain how a shark receives energy from the sun.

Snorkeling Activity

1) Snorkel quietly and determine which species that you observe and identify whether they are consumers or producers.

2) Identify at least one consumer and one producer on the reef and explain why you identify them as such.

4 SYMBIOTIC RELATIONSHIPS ON THE REEF

> **Lesson objectives**
> 1. Students will be able to identify the different types of symbiotic relationships
> 2. Students will observe fish behavior and draw conclusions regarding how the different species benefit from each other.

Biology Essential Concept:

Ecosystems are built around species relationships.

Snorkelers who are observant will notice how various species of fish tend to have relationships with other species. In some cases, the fish depend on each other for food and in other cases, they depend on other species for cleaning or protection. In the photo below, the Blue Tang is being 'cleaned' by the Cleaning Gobies who are eating the parasites off of the scales of the Tang. This benefits the Goby by providing it with a food source and helps the Tang by removing the parasites from the scales. This is an example of a **mutualistic** symbiotic relationship.

> **Vocabulary**
> **mutualistic** – symbiotic relationship which benefits both organisms in the relationship

Mutualistic Relationship

If the symbiotic relationship is to the benefit of both species, the symbiotic relationship is considered mutualistic. The example of the goby and the Blue

Tang is an example of mutualistic since both species receive benefit from the relationship.

Parasites

Parasitism is where the relationship benefits one species but is harmful to the other. In the photo, a Flamingo Tongue is on a Venus Fan coral. The mollusk (Flamingo Tongue) eats the polyps as it moves along the coral (the trail is visible in the photo).

The coral is scarred by the mollusk eating of the coral, therefore the Flamingo Tongue is considered a parasite.

Commensalism

In a **commensal** relationship, the relationship between the two species is characterized by one species receiving benefit and the other species is not affected in any way. In the photo below, a wrasses (the smaller fish) is following along behind a parrotfish as it feeds. The wrasses eats behind the parrotfish, taking advantage of the silt that is stirred up by the parrotfish. The wrasses gets the benefit of the food that is stirred up as well as protection from the larger species while the parrotfish receives no benefit at all.

Check for Understanding:

1) Describe the different types of symbiotic relationships that species have in an ecosystem.
2) Give examples of the different types of symbiotic relationships on the coral reef and in other ecosystems.

Snorkeling Activity

1) Snorkel quietly and watch various species interact and attempt to identify different relationships on the coral reef.

5 BIOLOGICAL CLASSIFICATION

> **Lesson objectives**
> 1. Students will be able to describe how organisms are classified and organized into various groups.
> 2. Students will be able to describe the basic differences between the major classifications of plants and animals.
> 3. Students will be able to describe the basic differences in animal and plant cells.

Biology Essential Concept:

Organisms are classified and organized according to their similarities.

With the vast variety of species throughout the world. It is important that a system for organizing the variety of species exists so that life forms can be studied efficiently. Since biology is the study of living organisms, an important concept is understanding how organisms are classified. This section will briefly describe how species are organized.

Later sections of the book will go into more specifics about the various families of animals and plants with this section focusing on the basic differences in the major classification levels.

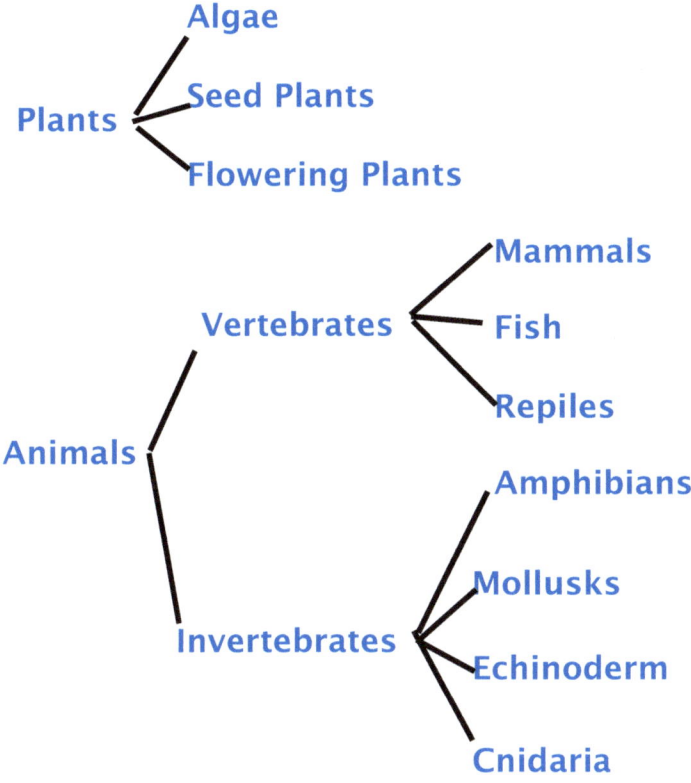

All living organisms can be classified by their characteristics. The diagram above shows an example of a very basic classification scheme, many more classifications exist. As the characteristics of plants and animals become more familiar, the classification system will become more clear. An important scientific concept is the understanding of how plants and animals are classified by their structure and other characteristics.

Characteristics of Plants:

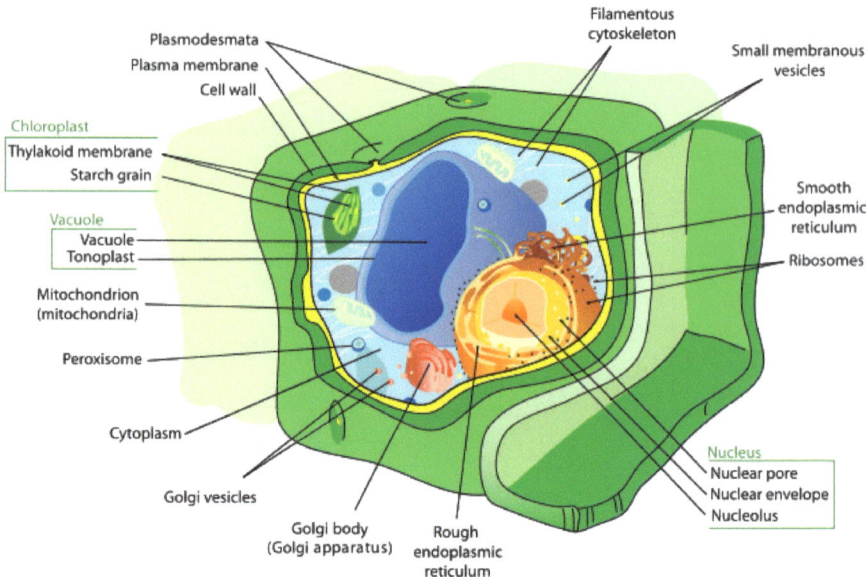

illustrations from public domain - Wikipedia

Plants have the following characteristics:
1) **Plants are multicellular.** Plants are comprised of multiple cells.

2) **Plants make their own food from the energy of the sun.** Through a process called **photosynthesis**, plants are able to convert the sun's energy into food. Within the plant cell, a chloroplast is present which makes food for the plant.

3) **The walls of a plant cell have a cell wall.** The presence of a cell wall in plants gives the plant a more rigid structure.

Characteristics of Animals

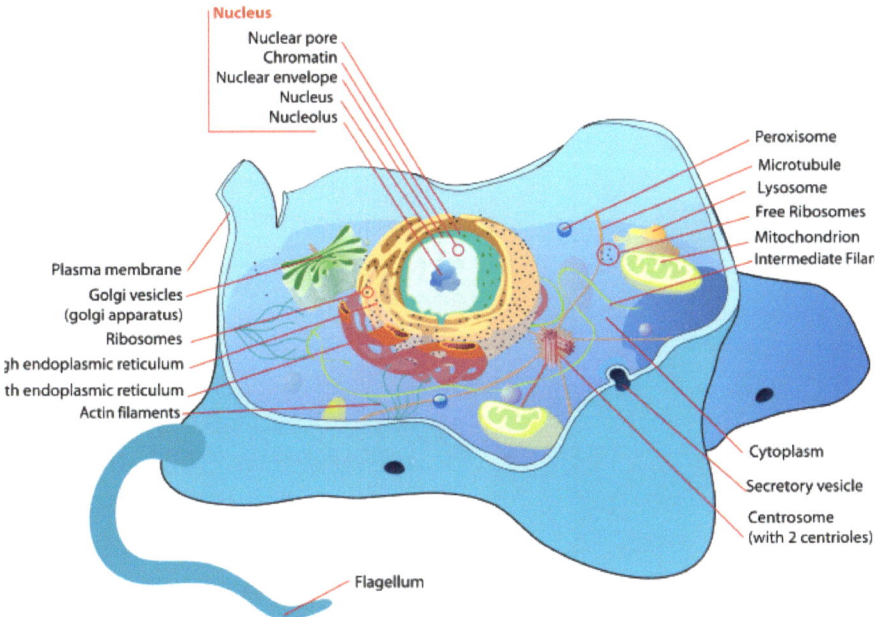

illustrations from public domain - Wikipedia

Generally, animals have the following characteristics:
1) **Animals are multicellular.** Animals are comprised of multiple cells.

2) **Animals take food from other organisms.** Animals must consume other organisms in order to obtain energy.

3) **Animal cells do not have a cell wall and do not have chloroplasts.**

MARINE ANIMALS

Vertebrates

One of the major classifications of animals are **vertebrates**. Vertebrates are distinguished by the fact that they have a backbone. All animals that have a backbone are classified into the **phylum** of chordata (vertebrate), referring to the spinal cord. More specific characteristics break the phylum of vertebrates into smaller classifications of **class**.

Bony fish is one of the large classes of vertebrates. It includes fish species which have bones (as opposed to cartilage like sharks). All bony fish have gills which provide them with the ability to filter oxygen from water. Most fish are **cold-blooded** meaning that their body temperature is controlled by the temperature of the water that they are in. They can not change their body temperature. Most fish lay eggs to reproduce.

Reptile is another class of vertebrates. One of the common reptiles that may be encountered on a coral reef is the sea turtle. Reptiles are **cold-blooded** animals. Reptiles lays their eggs on the land, even those that spend the majority of their life in the ocean.

Invertebrates

Invertebrates are those animals that do not have a spinal cord. They include the classes of echinoderms, mollusks, arthropods and others. For our study of the coral reef system, we will be most interested in the characteristics of **cnidaria**, **echinoderm** and **mollusks**.

Cnidaria is a marine animal phylum which includes coral, anemones and jellyfish. The distinguishing characteristic of the phylum is the specialized cells used to capture prey. The phylum is divided into swimming cnidaria (jellyfish) and **sessile** polyps (coral and anemones). Sessile refers to the fact that they can not move themselves, that are attached to a structure or the bottom of the ocean.

Echinoderm is a marine animal phylum which includes animals that are characterized by a five radial symmetry. It is the largest phylum that includes no animals outside the marine ecosystem. An example of an echinoderm is the starfish.

Mollusk is an animal phylum which includes invertebrate species that live in both fresh and salt water environments. Mollusks are divided into numerous classes including **cephalopods**, **gastropods**, and **bivalves**.

More specific information about the various phylums and the organisms that belong to the groups will be provided in later chapters.

MARINE PLANTS

Sea Grasses

Sea grasses are flowering plants. They only differ from land flowering plants by the fact that they live in the water. In all other ways, they are typical flowering plants with roots and stems. (Pictured above with a Trunkfish).

The roots of the sea grass provide a means to hold the grass in place but also to extract nutrients from the soil. Nutrients are extracted from the soil that they are rooted into and transport the nutrients through the plant via the **vascular system** of tubes that run through the plant.

Algae

Algae differ from sea grass in that it does not have roots, stems or a vascular system. Algae may contain 'hold-fasts' which are 'root-like' structures that connect the algae to a rock or other object. A hold-fast differs from a root because it does not extract nutrients from the soil or the object that it is connected to. Algae produces all of its food and, its entire structure is used for making food. This differs from sea grasses where only the leaves are used for food production. Algae do not have a vascular system, they must diffuse water in order to extract nutrients.

Algae include seaweed.

6- Vertebrates on the Reef – Anatomy

> **Learning Objectives**
>
> 1. Students will be able to make determinations about specific fish species by examining the anatomy.
>
> 2. Students will be able to make some determinations about the behaviors of the fish based on anatomy.

> **Biology Essential Concept:**
>
> Living organisms adapt to their environments.

Coral reefs are full of a great variety of vertebrates including fish, turtles, sharks and rays. A snorkeler will most certainly see many different types of fish, and a lucky snorkeler will see a sea turtle or ray. More enjoyment comes from getting to know the various species on the reef and observing their behaviors as they interact.

The most numerous vertebrate on the coral reef, by far, are fish. By noticing the specific anatomy of the various fish species on the coral reef, the snorkeler can make some determinations about the habits and characteristics of the species. Organisms change over time in order to obtain food and protect themselves.

Body Shape

Compare the species above. The French Angelfish (top) is a medium sized fish and has a flat, plate shaped body. The Bluehead Wrasses (middle) is smaller and more torpedo shaped. The barracuda (bottom left) is a larger fish and has a long thin body. Why would fish who live in the same ecosystem have such a different shape? The fish have different feeding habits and protection needs that require different types of movement and body shapes.

A flat body, like the angelfish, is more maneuverable in the crevices of the coral reef. The flat body allows the fish to make quick turns and changes in direction and the thin body allows it to slide in between the crevices of the coral to avoid predators. The flat, plate-like shape of the angelfish also makes it harder for predators to bite, therefore, offering some added protection. The rounded body of the wrasses allows it to move about very easily and move fast for longer periods of time. The barracuda has a very streamlined body that can move quickly in a straight line, but it does not change direction very easily at the faster speeds. It is much more capable of chasing down its prey due to the speed with which it travels.

Fin Anatomy

<u>Anal Fin</u> – Stabilizes the fish while moving through the water.

<u>Pelvic Fin</u> – Helps the fish move up and down in the water and making sharp and sudden turns.

Pectoral Fin – Helps the fish maintain depth in water. Some fish have specialized pectoral fins that can assist with 'walking' along the ocean floor and even flying.

Caudel Fin – (tail fin) Propels the fish through the water. The shape of the tail can give an indicatation as to how the fish moves.

Dorsal Fin – Holds the fish upright and keeps the fish from rolling over. Some fish have multiple dorsal fins.

Tail Shape– The different shapes of tail provide for specific swimming technique to avoid predators and chase down potential food. Forked tails are better for very fast speeds for long periods of time but do not provide for much quick turning. Rounded tail shapes allow the fish to make very quick turns and accelerate very quickly but is not suited to long distances. The different shapes of the tail provide different advantages to the fish.

Mouth Shape

Compare the species above. Notice how the shape and placement of the mouth is different on the examples. The mouth shape of a fish is a clue to how it gathers its food.

The fish to the lower left (a Butterfly fish) has a **terminal** mouth (pointing straight). The butterfly fish eats coral polyps and sea anemone so the elongated snout helps the fish reach into those places where the food source lives. When butterfly fish are seen while snorkeling, they are searching around the coral crevices poking around into the coral or along the bottom.

The fish to the right is a Parrotfish. It eats algae from the coral. It actually eats coral, digesting the algae that is growing on the coral and then expels the coral exoskeleton as sand. Because of how the Parrotfish eats, they have a 'beak-like' mouth with teeth that stick out in order to reach and bite off the coral. A quiet snorkeler can actually hear the Parrotfish eating as it bites on the coral.

The fish on the right, with the **superior** mouth, (Angelfish) eats on algae so its mouth is designed for reaching up under coral ledges on the algae hanging or clinging to the bottom of the ledge. When an angel fish is seen while snorkeling, they will usually be near the bottom and foraging around the ledges of the coral.

Eye Placement- The placement of the eyes on a fish gives a hint regarding where the fish finds its food. Eyes that are at the top of the head are found on fish that get their food on the top of the water. Typically, the eyes of a fish point in the direction in which they typically find their food.

Markings and Coloration- Fish generally have characteristics markings on their body. The markings are not just for decoration but provide either protection from or warnings to predators. The butterfly fish (to the right) has a stripe that runs through its eyes, hiding them, and has a large spots on the tail end which looks like eyes. These markings confuse a predator. When the predator approaches, the butterfly fish escapes in an unexpected direction and thus, eludes the predator.

Many fish in the reefs of the Caribbean change color as a means of disguise. Blue tangs will change from a very light blue to an almost black depending on the situation. Fish markings change as they move through the various stages of maturity. Some fish can change color quickly to signal to others or to confuse predators.

Check for Understanding

1) Describe the relationship of body shape to fish swimming behavior.
2) Describe the relationship of mouth placement to fish feeding behavior.
3) Describe the relationship of eye placement to fish feeding behavior.
4) Describe the function of various fins on a fish.
5) Describe the function of coloration and markings on a fish.
6) Describe how fish have adapted over time to meet their feeding and protection needs.

Snorkeling Activity

1) Select two fish that are common in the area that you are snorkeling, describe each one by coloration, body shape, tail shape, mouth placement, eye placement and fins.
2) Describe the predicted behaviors of the two fish you selected based on anatomy. Describe how the anatomy helps them gather food and avoid predators.

7 VERTEBRATES ON THE REEF – BEHAVIOR

> **Lesson Objectives**
> 1. Students will be able to identify how various species on the reef behave and interact with each other.

Swimming Behavior

While snorkeling in the coral reef areas of the Caribbean, particular patterns of behavior of the marine species will become apparent to the careful observer. While swimming 'quietly' (floating without much movement), the marine species will relax and begin to 'go about' their normal business. As long as they do not feel threatened, they will fall into their normal routine. Pay close attention and consider how the various species of the system interact with each other. A great deal can be learned about the marine species by watching their behavior and drawing some inferences about why they act as they do. After a couple visits to snorkeling spots, the common species will begin to become familiar and specific patterns of behavior will be noticed.

How much do the species roam through the reef area?

It will be noticed that some fish seem to swim around without any regard to a specific 'home' within the coral reef, while others, stay very close to a particular spot. Some of those species are actually quite assertive in attempts to keep other fish away from their area. The damselfish, although sometimes very small, will attempt to chase away any intruder (including a snorkeler) near their home. They guard their home without fail. They will dart around a specific spot and will attempt to scare away anything that comes close.

The Parrotfish roams around chomping on coral. They do not tend to stay within any specific area, but much like many other fish on the reef, they tend to venture around constantly.

A careful observer will notice that some fish can be found repeatedly around the same rock or spot even on repeat visits on different days.

How much do the species interact with other species?

An observant swimmer will sometimes see fish who seem to swim with other types of fish and follow along with them. As this behavior is seen, consider why they seem to hang around together (there is almost always a reason).

Where do the different species tend to be seen?

Snorkelers will notice that some fish are always found hiding in the shadows, under a ledge of coral or crevice. It will be noticed that some fish tend to swim very shallow and are always seen very near the surface of the water. Some fish will be located 'hanging out' in the grassy areas while others are swimming around a particular type of coral or buried in the sand. Observe the fish behaviors and try to determine what advantages the fish may have with where they tend to stay.

How many of the species tend to travel together?

Some marine species tend to travel alone and others tend to travel in groups. Occasionally a blue tang (below) will swim alone, however, usually they are in very large groups swimming and feeding together.

Snorkeling Activity

1) While snorkeling, notice how the various species are moving about the reef and it they tend to be in groups or as singles? If they are moving as groups, make a prediction as to what benefit they receive by moving as a group. Notice how other species 'interact' with the group.

8 – Common Vertebrates on the Reef – Identification

A great variety of fish are found on the reef. A casual snorkeler will have more enjoyment when they can identify the various species. Although many species have variations due to camouflage techniques and the various stages of development, many of the species are identifiable. A snorkeler can expect to see many of the examples listed here in abundance.

> **Learning Objectives**
>
> 1. Students will be able to identify the very common fish on the reef.

Banded Butterflyfish

Spotfin Butterflyfish with a Schoolmaster

Bluestriped Grunt

Beaugregory (juvenile)
Rather small fish - usually seen in the grasses.

French Angelfish

Sargent Major

Foureye Butterflyfish

Squirrelfish

Trumpetfish

Blue Tang
To the right - an Intermediate phase Blue Tang

Lizardfish

Yellowtail Damselfish

Smallmouth Grunt

STING RAYS

> **Learning Objectives**
>
> 1. Students will understand the relationship between rays and sharks.

Sting Rays are found throughout the Caribbean and Florida. They are usually found in the sandy areas and tend to hover around the bottom. They will, when resting, bury down into the sand, making them somewhat difficult to see. Although they can cause injury to swimmers with the stinging barb on their tail, they are shy of humans and are not aggressive. Injuries that result from sting rays occur when a swimmer steps directly on a resting ray. Because the use of the barb usually results in the death of the ray, they only use the barb as a last resort. By being careful while walking through the water, noticing the bottom and shuffling their feet, swimmers can almost certainly avoid the sting of the ray.

Sting Ray Anatomy

Rays are batoideas and closely related to the shark. They do not have boney skeletons, but their skeletons are made of strong cartilage. They are distinguished from sharks by their distinctive flat, winged bodies. The winged appendages are actually their enlarged pectoral fins that are fused into their head. Additionally, they have gills slits on their bottom side.

Stinger – On the end of the long tail is a venomous barb. If danger is sensed by the ray, they will raise the tail quickly to a vertical position and 'slap' the barb into the aggressor. In the case of a swimmer, the barb usually embeds into the calf. Depending on the force with which the barb enters the skin, it sometimes has to be cut from the leg. The injury is extremely painful. A stingray which 'deploys' its barb dies within hours of the attack. Because the use of the barbed stinger results in injury or death of the ray, the ray only uses the defense when it feels extremely threatened.

Sting Ray Feeding Behavior

Sting rays feed on crustaceans and small fish in the sand. They flutter through the sand when trying to find their food. Because of the location of their eyes on top of their flat body with the mouth below, they can not see their prey while moving through the sand. Therefore, they must rely on the sensitive smelling sense to identify their food in the sand.

Spotted Eagle Ray

Check for Understanding
1) Describe the similarities between rays and sharks.
2) Describe how to avoid being stung by a sting ray.
3) Describe how rays eat.

9- INVERTEBRATES ON THE REEF

The coral reef is full of invertebrates. Some are familiar, some are not. A snorkeler on the reef who is a careful observer can see a huge variety of invertebrate specimens.

Invertebrates include coral, echinoderms, mollusks, worms, squid and many others. This text will focus on the most common and easily identified.

Lesson Objectives

1. Students will understand the basic body structure of a coral.

2. Students will understand the process of coral growth in the reef development.

3. Students will be able to identify the various types of common invertebrates that live on the coral reef.

Coral

Coral belongs to the cnidaria phylum (pronounced nye-dare-ee-a) and is a system of many invertebrate **polyps** that share a common vascular canal that allow the individual polyps to transfer nutrients throughout the system. The polyps excrete a calcium carbonate that, over time, produces the coral exoskeleton. The exoskeleton is what most people 'know' as coral. The hard substance that most call 'coral' is actually this substance that is produced by the animal that lives inside.

Coral Anatomy and Feeding Behavior

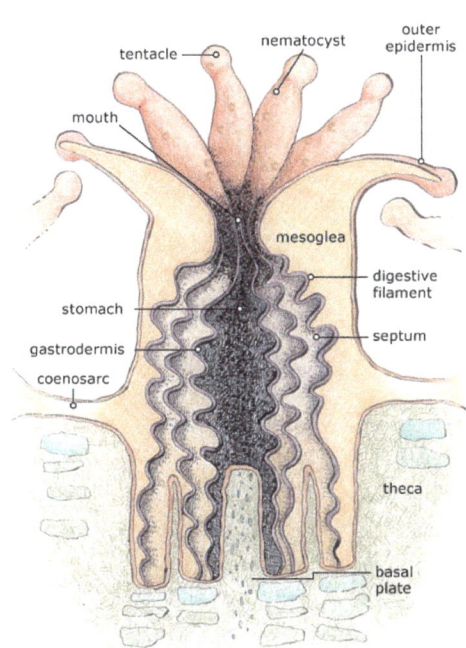

illustration from Wikipedia-public

The illustration to the left shows the anatomy of the coral polyp.

The exoskeleton of the coral is produced at the bottom of the stomach and formed by the basal plates. Over generations the basal plates stack which 'grows' the coral.

The coral head, which is made up of a colony of coral polyps, lives within the exoskeleton portion of the coral that is most familiar. The exoskeleton provides the protection to the animal living inside. When the polyp feels stressed, it sinks back into the protection of the exoskeleton. Because the polyps are easily stressed, a daytime snorkeler will only catch glimpses of the coral polyps before it sinks into hiding. At night, the coral is feeding and is most easily visible.

> **Vocabulary**
> **polyp** – single organism that is cylindrical in shape which live with others in a colony

Feeding – The polyps eat mostly plankton or, in some cases, small fish or other small organisms. The tentacles have nematocysts which contain poison to kill the prey. The tentacles pull the prey into the stomach of the polyp and then the polyp digests the nutrients from the prey and pushes the remains back out the top of the polyp into the water.

Snorkeling Activity
1) While snorkeling, notice the structures of various coral species on the reef.
2) While snorkeling, attempt to identify the specific species of coral listed on the following pages.

Check for Understanding
1) Describe how coral polyps eat and expel waste.
2) Describe how coral grows the exoskeleton that protects it.

Star Coral

(with some Christmas Tree worms)

Fire Coral

(Will burn the skin with contact)

Black Sea Rod Coral

Pillar Coral

Venus Sea Fan

Elkhorn Coral

SUITCASE STUDIES

Echinoderms

Echinoderms are marine animals whose body shape is distinctive. They are symmetrical and typically have appendages which radiate from the center. Examples are starfish, sea urchins and sand dollars.

Echinoderms have the distinction of being able to regenerate lost appendages. If a starfish has an arm broken off, they can grow a new one in its place. Snorkelers in the Caribbean will certainly see sea urchins hiding in the crevices of the coral and rocks.

Sea Urchins

Sea urchins are some of the most plentiful echinoderm recognized on the coral reef. They are not obvious to swimmers, however, by looking closely in the nooks and crevices of rocks or coral shelfs, they will certainly become apparent.

Sea urchins, unlike starfish, are 'global' in structure and do not have arms projecting from their disc. The disc, if turned over, has five parts radiating from the center. They have a mouth opening in the center of the disc surrounded by soft lips with small embedded hard structures used like teeth. Around the mouth are five sets of tube feet and five sets of gills.

While it may appear that the sea urchins do not move, they actually do move, slowly, like the starfish with tubular feet on their disc. Unlike the starfish, the sea urchin can not move its appendages. The only movement that the sea urchin can exhibit is the movement across a structure with its tubular feet. Because of this, the spines are essential for protection.

The sea urchins that are the most plentiful in the Caribbean have long, dark spines extending from the body. These spines are painful if they penetrate the skin and some may be slightly venomous. They are not considered poisonous to humans.

Sea Cucumber

The sea cucumber is one of the most interesting echinoderms in the ocean. Although closely related to the starfish and sea urchin, it shares little similarity in its appearance.

The sea cucumber scavenges its meals as it moves across the ocean floor. It is **benthic**, meaning that it lives on the ocean floor. It eats almost anything. Like other echinoderms, it has five section divisions in its body. The body shape is elongated with the mouth disc located at one end.

One of the interesting facts about sea cucumbers is that it has the ability to change the consistency of its skin so that it can be hard or take on the characteristic of a liquid or gel. This allows it to slip into the smallest of crevices when necessary. Another interesting fact about sea cucumbers is that they have no brain. It does have sensory fibers that run through the body to allow it to sense some aspects of its surroundings, but there does not appear to be a central system that controls its functions. Also, some sea cucumbers can expel a sticky membranes from its **anus** that can capture prey or scare off predators.

Because of the manner in which the sea cucumbers clean the ocean floor, one can often find a sea cucumber where tubes of waste product are seen lying on the bottom of the ocean.

Starfish

Starfish are some of the most easily recognizable echinoderms on the reef. They are distinguished by the central disc and the five radiating legs (some species have more than five legs). The top surface of the starfish may be smooth or rough with some species having very spiny surfaces. They are found in a variety of colors.

All starfish have tubular feet and their mouth is located on the underside. Their eyes are also located at the end of their arms. Some species have light sensitive cells that can detect light even when their eyespots are covered. This assists the starfish in identifying predators.

The starfish are benthic predators of the reef ecosystem. They feed mostly on invertebrates and are, therefore, secondary consumers.

As is characteristic of most echinoderms, they can regenerate damaged parts or lost arms. Some species can grow and entire new body from only a portion of a remaining arm. Additionally, starfish can shed arms voluntarily, as a means of defense.

Starfish are **keystone species** of some ecosystems. A keystone species is one which has an unusually significant effect on the environment when compared to its number within the ecosystem. A few starfish, as a keystone species, has great effect on the balance of the system. If starfish, even a few, were removed from an environment, the species that it feeds on would grow out of control and therefore have detrimental effects on the balance of the ecosystem.

> **Check for Understanding**
> 1) Describe the anatomical characteristics of an echinoderm.
> 2) Describe a 'keystone' species.
> 3) Give examples of 'benthic' animals.

> **Snorkeling Activity**
> 1) Snorkel out to a coral reef spot and make an effort to peer into the crevices and holes among the coral. Attempt to identify the echinoderms that are buried in the crevices.
> 2) While snorkeling in open water, be on the look out for what appears like tubes of sand laying on the ocean floor. Many times, where these tubes are found, close by will be a sea cucumber. It will often appear to be a piece of wood or other inanimate object.

Mollusks

Mollusks are the group of invertebrates that include snails and slugs (gastropods), clams and oysters (bivalves) and cephalopods (squid and octopus). Mollusks are the most diverse and numerous of animals on earth. They are found in both fresh water and marine habitats and include a great variety of animal characteristics.

Visitors to coral reefs will most certainly see mollusks during a snorkel through the reef.

Common types of mollusks:

Cephalopods- The most advanced mollusks, include squid and octopus.

Gastropods– The most numerous type of mollusks, include snails and slugs. Gastropods make up 80% of the species included in mollusks.

Bivalve – The mollusks that have an upper and lower shell. This type of mollusk include the clam and oyster.

Characteristics of Mollusks:

Mollusks have a great variety of body structure, however, all mollusks have a **mantle**. The mantle is the cavity or space used for breathing and excretion of waste.

Those mollusks with a shell develop the shell by secreting proteins and **chitin** from the exterior of their mantle. Chitin is a complex substance formed from glucose sugars. Chitin is produced by animals which have hard exoskeletons like lobsters, crabs and some insects. Most squid and octopuses have an internal shell of some sort made from chitin.

Mollusks also have the distinction of using organs for multiple purposes. For example, the heart and kidney are important in the circulatory system and excretory system as well as for reproduction. In bivalves, the gills are used for breathing and to produce a current in the mantle cavity that is useful for excretion and reproduction.

Commonly Identified Reef Mollusks

Snails

Snails are types of gastropod mollusks that are relatively common in the Caribbean. They are characterized by their egg-shaped shells and flat bottoms. The Flamingo Tongue is often seen on Venus Fan Coral. The colorful exterior of the Flamingo Tongue is actually the animal's mantel which is spread over the shell.

Flamingo Tongue

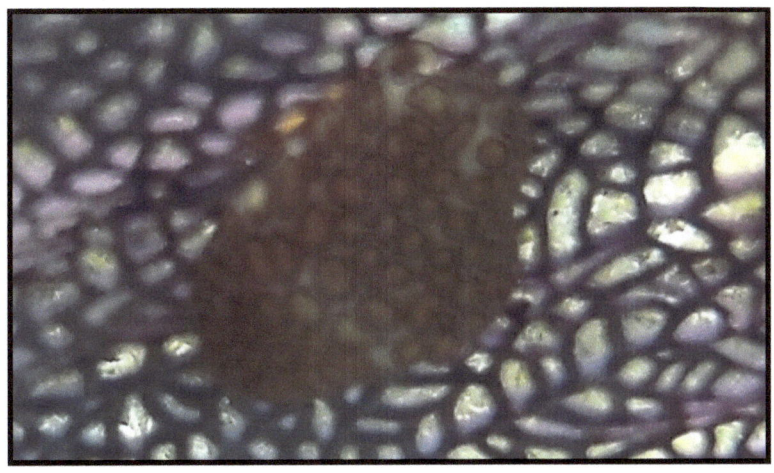

Squid

The Caribbean Reef squid is common around the reef areas of the Caribbean. They are generally found near the grassy areas located around a coral area. Generally the squid are seen in small groups swimming at mid-depth of the water. The squid are very interesting animals to observe as they attempt to scare off the snorkeler with their antics of twisting their tentacles into antler shaped appendages. The large eyes of the squid are intense as they 'stare down' the watchful snorkeler.

Body Parts of the Squid

Mantel– The mantel is the main body mass of the squid. Within the mantel are most of the internal organs. The mantel opens up to the excretory and digestive system at the bottom where the arms and tentacles come to the body. Also, at the base of the mantel is the siphon, (through which the squid pushes water in order to jet through the water) and the gills which are used to take oxygen from the water.

Arms and Tentacles – The Caribbean reef squid has eight arms and two tentacles. The tentacles are longer than the arms and have suction cups used to hold prey and pull it into the mouth located at the base of the mantle.

Nervous System – Squid and other cephalopods have the most complex nervous system of all invertebrates. Its brain is, by far, the largest. The brain is protected by a cartilage cranium.

Vision – When snorkeling, the swimmer will notice that the eyes of the squid are distinctive. They seem to be looking directly at the swimmer in a way that makes one consider what they may be thinking.

Squids have very good vision. Their vision plays a role in their ability to change colors to camouflage themselves from predators.

Camouflage – Squid and octopus have specials cells (chromatophores) to change the brightness and pattern of their skin to match the background that they see. A snorkeler who follows and watches squid will see them change colors often as they move.

Ink – All cephalopods have ink sacs which contain ink that can be expelled in a cloud to confuse predators.

Check for Understanding

1) Identify the parts of a mollusk.
2) Describe how molluks protect themselves from predators.
3) Identify the three types of mollusks and the characteristics of each.

Sponges

Sponges are animals that belong to the phylum of **porifera**. Sponges are distinguished by the holes that run through the body to allow water to flow throughout the organism. They do not have digestive, circulatory, or nervous systems but rely on the constant flow of water to provide it with nutrients and to eliminate waste.

Characteristics of Sponges:

Like all animals, sponges are multicellular and their cells do not have cell walls. The body of a sponge, unlike most animals, does not have **symmetry**. They are dependent on water's movement throughout the animal. As the water moves through the body, the sponge takes the nutrients from the water and excretes its waste. Sponges can be distinguished from coral by the presence of the pores or vessel-type openings.

Snorkeling Activity

1) Snorkel out and identify some sponges. Notice the large holes that are characteristic.

Check for Understanding

1) Identify the characteristics of a sponge.
2) Describe how the sponge supplies itself with nutrients.
3) Identify how the sponge eliminates waste products from its body.

Worms

A great variety of marine worms exist in the coral reef ecosystem. A careful observer can learn to identify them with a little practice. Many of these worms do not look like the typical terrestrial worm and may be misidentified or failed to be noticed. The structures of the marine worms are unique and specially adapted to the marine environment

Feather Duster

The Feather Duster is a worm that lives inside a tube that is built from sand, silt and parchment. The worm extends a feathery crown that is used to filter water for food. The worm is able to withdraw the feeding feathers into the tube when danger is near.

A snorkeler who is not aware of the presence of the Feather Duster worm, will mistake the worm appendages for a plant.

Social Feather Duster

The Social Feather Duster is very similar to the Feather Duster worm with the main distinction being that the Social Feather Dusters live in groups.

Christmas Tree Worms

Christmas Tree Worms are named for the unique shape of the gills of the worm. The worm extends its respiratory organs (gills) outside the tube in which it lives in order to filter the oxygen from the water like a fish's gills. The structures of the worm also serves as the feeding mechanisms which traps its food from the flow of the water.

    The worm lives inside a tube buried into coral or rock and do not have a mechanism for movement. It must rely on food that passes by and is filtered by the 'tree' structure. When danger is detected, the worm can withdraw into the tube for protection and virtually disappear.

Split Crown Feather Duster

Much like the Feather Duster, the Split Crown (above) has less filtering structures and can come in a variety of colors. They can be found in the crevices of coral.

Flatworms

Flatworms tend to stay on the bottom of the ocean or under ledges of coral. They are usually not a threat to humans, but if contact is made, they will burn the skin.

Sea Anemones

A sea anemone belongs to the phylum cnidaria which makes it a close relative to the coral. The sea anemone is a polyp animal, but differs from the coral due to the lack of the exoskeleton. The sea anemone polyps are larger than the polyps of most coral due to the fact that the anemone is a predator of larger animals.

Sea Anemone are usually attached to the floor of the ocean making them a **sessile** species, however some species do have a mechanism which allows them to float. The sea anemone polyp attaches itself to the ocean floor by a basal disc and has a set of tentacles which contain venom. The sea anemone is predatory and will eat whatever small fish or other animal that may venture within reach of the poisonous tentacles.

Sea Anemone are generally found on the reefs of the Caribbean in crevices and are not overly abundant but can be found by the observant snorkeler.

Check for Understanding

1) What is the primary defense mechanism of the tube worms like Feather Dusters and Christmas Tree worms?
2) How do sea anemone prey on other species?

Snorkeling Activity

1) Snorkel out and look for and identify the common tube worms (Christmas Tree worms and Feather Duster). When located, approach the worm and watch it disappear into its tube.

10 – PLANTS ON AND AROUND THE REEF

The plants on and around the coral reef are as diverse as the animals. A careful observer will be able to identify the various types of plants.

Mangroves

A mangrove swamp is a unique ecosystem characteristic of a swamp area covered by trees that grow from coastal sediments. They are found generally around the equator and extending both north and south about 25 degrees. Mangrove swamps are found in tropical and subtropical tidal areas along marine shorelines and **estuaries**.

> **Lesson Objectives**
>
> 1. Students will be able to describe the unique characteristics of a mangrove.
>
> 2. Students will be able to explain the importance of the mangrove to the ocean ecosystem.
>
> 3. The students will be able to identify some of the ways that the plants in the mangrove swamp adapt to the environment.

Mangroves are common throughout the Caribbean and are vital to the marine ecosystem and environmental systems within the coral reef.

Mangrove swamps provide rich habitat for specialized species of both **flora** and **fauna** .

Characteristics of a Mangrove

A mangrove is a particular species which has adapted to the special habitat. About 100 different species of mangroves have been identified, but typically, any vegetation which grows under the specialized circumstances are called 'mangroves'.

A mangrove swamp is made up of trees or shrubs which grow in the region of a body of saltwater. When the tide comes in, it floods the area with salt water making it necessary for the plant species to be able to withstand the saltwater. Additionally, when the tide goes out and the water depth becomes shallow, the **salinity** of the water increases to even higher levels. Then, when tide comes back in, the salinity again changes, therefore, it is necessary that species in the mangrove can adapt to ever-changing salinity levels.

In addition to the salinity of the area, the plants and animals must be able to withstand a constantly changing water temperatures. As the tides comes in bringing in water, the temperature of the water lowers. As the tide goes out and the water depth becomes lower, the sun heats up the water.

As a results of the constantly changing factors within the mangrove system, the species that live there must be adaptable to the conditions.

> **Vocabulary**
> **salinity** – level of dissolved salt level in the water
> **sediment** – floating matter that settles on the bottom

Importance of a Mangrove

Mangroves provide a structure which protects the shoreline. The trees and shrubs in the water where it washes in and out have developed very extensive root systems to hold them firmly into the soil. The roots provide structure for the **sediment** washed in from the tides to settle. This sediment structure provides protection against storm surges that would otherwise erode the shoreline.

The mangrove vegetation also slows the wave energy coming in from the tide providing for the sediment of heavy metals into the soil. These heavy metals gather and are, essentially, sifted from the ocean water.

Because these heavy metals, in concentration, are damaging to coral reefs and other marine ecosystems, the mangroves help by providing a deposit for these substances. However, this concentration of the heavy metals makes any disturbance of the mangroves particularly harmful to the surrounding ecosystems. When the mangrove plant species are disturbed or removed, the sediments are released in great concentration.

The intricate structures of the mangrove roots also provide habitat for a large variety of marine animals and plants. Because of the calm water within the mangrove area, many species use the areas for the depositing of eggs and their immature offspring. Additionally, some species prefer the calm area for the permanent habitat.

Mangrove Tree Specialization

The plants in the mangrove swamp have developed very specialized properties that allow them to thrive in the special habitat.

In order to survive the lack of the oxygen available in the water source, the plants have specialized roots which grow up above the soil and sediment in order to access the air. As a result, when one ventures into the mangrove swamps to explore, the trees tend to look like they are perching above the water on the structure of entangled roots.

In order to remove the excessive salt from the water, some species are able to 'send' salt to the leaves and then excrete the salt in order to preserve a healthy level of salt content. (see the picture to the left – salt crystals on a leaf).

When the plant is needing to preserve the level of fresh water within their system, the plant can regulate the size of their **stomata** (pores on the leave that is used to control substances going in and out of the leaf). By constricting the stomata, the plant can store the fresh water within the plant as needed.

Plants in the mangrove swamp have a special need to be efficient with the lack of nutrients that are available. The plants in the mangrove can store healthy gases in their roots and process the nutrients within the roots as needed and even when under water. Plants in more typical environments must have a constant source of nutrients through their root system.

A final specialization comes from how the plants insure the survival of their offspring. Many of the plants in the mangrove swamp produce seeds which float. This allows the seeds to wash ashore in a suitable location to take root. A walk along the beach near a mangrove swamp will find seeds along the water's edge that have floated from the mangrove. Another method of survival involves the parent plant providing a site for the seed to remain until it germinates. Depending on the species, it may  germinate while inside a fruit or may grow through the fruit, using the fruit as a source of nutrients. The parent plant will drop the germinated seed and it will float continuing to grow.

Check for Understanding

1) How do mangrove trees adapt to the extreme environment?
2) Why are mangroves important?

Sea Grasses

Characteristics of Sea Grasses

Sea grasses are flowering plants that are distinguished by the fact that they have long slender leaves. They pollinate the flowers through the water.

Because sea grasses depend on photosynthesis for their food production, they are only found in water that is shallow enough for the light penetration. Water that is shallow enough for photosynthesis is known as the **photic zone**.

Lesson Objectives

1. Students will be able to describe the difference in marine plants and algae.

2. Students will be able to describe the basics of the process of photosynthesis.

3. Students will be able to explain the importance of sea grasses and algae to the coral reef ecosystem.

Importance of Sea Grasses

Sea grasses provide an important habitat for juvenile fish types in addition to mollusks, nematodes and many others.

Sea grasses are important producers on the reef. From the energy gathered by the plants from the sun, photosynthesis allows the plants to convert that energy into food. When the animals on the reef consume the grasses, that energy from the sun is transferred to the animal. Sea grasses provide food for some animals such as sea turtles, sea urchins and crabs.

Sea grasses are also important in the creation of the reef structures. The grasses provide some resistance to the water as it moves in due to tides or the wave energy and it slows the force of the water, reducing the amount of sediment in the water. This reduction in the sediment is important to a healthy coral system. Additionally, the slowing of the force of the waves, protects the shore line from erosion.

Algae

Algae is a varied type of organism which makes its own food through **photosynthesis**. Algae can range from a single cell to a very complex organism such as seaweed or kelp. In the Caribbean, various types of algae will be seen by the casual snorkeler.

How are Algae different from Water Plants?

Algae can range from very simple organisms to very complex species such as kelp and seaweed. Seaweeds technically are algae, even though they are rather complex. Algae typically differ from plants in that plants have specialized cell types where algae, generally, do not. Where plants have roots, stems, leaves and other specialized parts, algae does not.

Photosynthesis Process

Photosynthesis is the process by which plants take energy from the sun and convert it to energy that the organism can use for its cellular processes. Simply put, photosynthesis occurs when light energy combined with water and carbon dioxide creates a reaction within the chloroplasts of the plant cells. The reaction creates sugars which is used by the plant as energy. A by-product of the process is oxygen.

Algae in the Food Chain of the Reef

Algae is one of the **producers** on the reef. Because it can make its own food through photosynthesis and other species feed on the algae, it is considered at the producer level of the food web. As a result, algae are extremely important to the reef ecosystem. Without algae, many other organisms which depend on the algae for food would not exist on the reef ecosystem. A snorkeler on the coral reef will notice many fish species eating algae.

Common Types of Algae

Brown Algae Examples

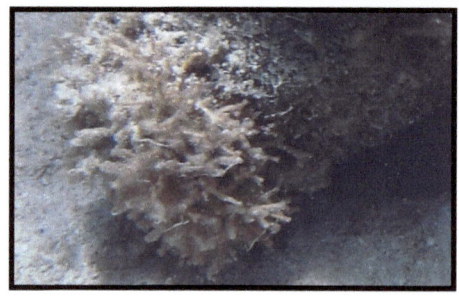

Green Algae Examples

Check for Understanding

1) Describe how algae and marine plants differ.
2) Describe the process of photosynthesis.
3) Explain why sea grasses are important to the coral reef ecosystem.
4) Explain why marine plants do not grow in the very deep areas of the ocean.

ABOUT THE AUTHOR

Dawn Lamb is a retired educator who spends time traveling, particularly in the Caribbean. The passion for education continues as she seeks to create educational resources for families who want to provided learning experiences in various places in the world. She can be reached at dlamb@suitcasestudies.com.

www.ingramcontent.com/pod-product-compliance
Lightning Source LLC
Chambersburg PA
CBHW040906180526
45159CB00010BA/2942